The Fourth Wave

Evolutionary Technology to Revolutionize Money Making

Deepthi Ayyagari

WordBuzz Publishing
(Indie-published by author-publisher)
Copyright © 2017 by Deepthi Ayyagari
All Rights Reserved.
First published: August, 2017.

The author-publisher, Deepthi Ayyagari asserts the moral right to be identified as the owner of this book. No part of this book may be reproduced, stored in a retrieval system, transmitted, recorded or copied in any format, mechanical or electronic, without the explicit written permission of the author-publisher.

Front cover image courtesy: Samuel Zeller

Disclaimer:
All the material presented in this book is purely for entertainment purposes, and not to serve as expert, professional advice. The author-publisher is not liable, under any circumstances, for any loss or damage, directly or indirectly resulting out of the usage of the material in this work. Readers are strictly advised to consume this work for their entertainment purposes alone, and any implementation for business purposes or otherwise, be taken up based on expert professional advice. While every attempt has been made to ensure that the content is error-free, the author-publisher does not take responsibility for the accuracy and correctness of information and content provided in this book, including factual and statistical data. Any reference to trademarks and their product and service offerings is strictly to evoke the reader's interest. All the trademarks are the property of their respective owners, and any reference to individuals, services, products or companies is not for endorsement purposes, and the use of all information, including, but not limited to the aforesaid from this work is at the reader's own risk.

To my dearest Hari Chandana...

The advent of the fourth wave of technology–virtual, augmented and mixed reality, is dramatically changing the way we perceive and do things. Today, this technology has opened a world of possibilities in ways we could not have fathomed a decade earlier. As cryptocurrencies are forging their way into replacing the traditional currencies of the world, we can see that virtual realism is the new world-order. There is no area left untouched by this fourth wave–medicine, robotics, engineering, military, sports and gaming–every area known to man is under an express transformation.

The financial services industry needs to catch up with newer ways of doing business and presenting innovative solutions that will add value to clients and share-holders. The value that technically innovative solutions present in terms of ROI just makes it impossible to ignore the possibilities that exist today. Because what we do today, will shape our tomorrow. This book delves on the business of investment and wealth management from a scientific as well as an artistic perspective, through deploying technology to our advantage and unleashing the power of perceptual experience. It is finally left to us to either ride the wave or be drowned under its current. But the wave is already here!

About the author:

Deepthi Ayyagari has a decade and a half's experience in the software technology industry in delivering products that have immense value for clients. After dabbing her hands in the domains of insurance and health care, amongst others, she is working with a financial services industry major for the past one decade.

She believes that adding value to what you offer clients should be an everyday business. For in the end, two things matter–what business value your products deliver to your clients, and what business value your clients derive from your products!

Author's Note

This book is for all the technology, business, and idea enthusiasts in the world who are passionate about adding value to people's lives. It is for all of those who are inquisitive about how our world is set to be transformed in ways we can be as creative to think as we please. This is not an overstatement at all, as we will see through this book. Although a lot of this book is about the Wall Street business, it will give the reader a taste of the fourth wave of technology, and how it is impacting every area of our lives. If you want to pick this book, you need not be a technical or financial geek; it will do if you are just plain inquisitive about the way things are changing through innovative technologies. This world is a world of possibilities, and this book is merely a window into it…

For businesses, ideas in the book would demonstrate how to empower your end customer. The simple reason, is this- in business, the goal should not be left to empowering your own systems and processes, products and solutions. An intelligent business needs to empower its customers so that they can, in turn, empower the business by dealing with it on a time and again basis. It takes the reader into the power of perceptual experience- how you can win clients by getting them to taste your product through virtual, augmented or mixed reality. The individual use cases may be easily used as blueprints to power any business with the needed customization.

For individuals, this book presents several interesting and innovative applications that are made possible through the implementation of disruptive technologies, including the fourth wave. The dimension that it presents for a positive change in customer experience is certainly worth looking forward to.

Contents

1) Perception Is Everything!
2) Technological Trends That Are Redefining the Future
3) How the Fourth Wave of Technology Is Changing Businesses
4) Investments and Wealth Management at the Cusp of Change
5) The Age of Millennials and Centennials
6) Products That Have Scaled Up the Innovation Ladder
7) Innovative Products Through the Fourth Wave–A World of Opportunities
8) Watching the 'Wave'
9) Conclusion

Perception Is Everything!

What would you say if you suddenly happened to see colors that you never saw before? How much money would you spend on a pair of sunglasses that would enable you to see the world differently, and more vividly than you have ever seen it?

A scientist, Mr. Don McPherson has developed glasses that literally change one's perception of the world. Quirkily enough, he happened to come to know of his own discovery back in the year 2005. By the year 2010, he has come to co-found a company that is changing the way many people look at the world today.

EnChroma Inc., based in Berkeley, California, is reaping profits out of being able to restore color perception to those suffering from color blindness. Even if one does not suffer from the medical condition, the glasses enhance color perception to an entirely new level, thus making them irresistible for anyone using them.

These glasses were originally intended to aid surgeons through laser surgery for protecting their eyes from the harmful effects of laser. For this purpose, McPherson, a Ph.D. holder in glass science, had experimented with using some rare minerals to make lens that not only protects their eyes, but also enables doctors to be able to distinctively see blood and tissues as they operated upon their patients. Since they were found to provide a vivid picture with an incredible amount of saturation of colors, the glasses became a favorite accessory that surgeons loved to carry everywhere they went. It is no wonder that the glasses were no longer to be found in operating theatres alone. McPherson himself started wearing them. One day, a friend of his, one Mr. Angell, who was color blind, happened to borrow his pair of glasses during a Frisbee game, and was pleasantly startled when they enabled him to see the orange color for the first time. This gave a kick-start to the business that EnChroma is doing, and doing well today.

It is estimated that some 300 million people are affected by being color blind the world over. This medical condition is a result of one or more of the three cones in the eyes being faulty, so they would not be able to signal the right color to the brain. These lenses have a mechanism to filter out some wave lengths of light to distinctively see colors, with the result that most people affected by color blindness now have a solution to their problem.

A mathematician named Schmeder created a model of color vision by incorporating 10,000 colors, natural as well as man-made. This, he imbibed into the design of the eyewear to further boost the effect of the lens.

The company offers glasses for not just the color blind, but to anyone who loves to see things better. They offer color enhancing glasses with UV-ray protection in a wide array of frames, and also options from which to choose the type of lens.

Coming to the point, who does not want to see things better? This is exactly the point of using virtual, augmented and mixed reality. Perception is everything! It is the core differentiator of your business from that of your next-door vendor.

In the industry of financial investment services, it is of pivotal importance that the potential investors 'see' how they are going to benefit from their investments. The market is literally flooded with market leaders. There is a wide array of choices available, which meet the same business goal that so many specialized product vendors, investment banks and brokerage houses have on offer.

The point is, it is finally the market makers rather than the market leaders that emerge as the winners of the game. Market makers–is in the sense that they are those disruptive innovators that create the market for a new product or service offering.

Now let us face the fact–the fact is that there is nothing very innovative about using virtual/augmented/mixed reality now. The current scenario is that so many products using these technologies are already being widely used by customers the world over. It is just that the financial services industry is yet to adopt new technology and adapt to the changing scene. How fast one catches up with technological trends and uses them for utmost business advantage alone matters the most. Facts be spoken of, there are several companies that have already jumped onto the bandwagon for a good ride in this gold mine, and nobody would want to be left out.

In this field abounding with implementing complex algorithms and math formulae to derive anything from getting the right portfolio for investment to providing a statement of returns, it only makes sense to present complex products in a simple but innovative manner.
When customers see some tangibility in the returns that they can possibly get in the future, they might likely invest more. For many customers, how and how well you predict the year on year returns in wealth management products is not a factor at all. What matters is how you can enable them to pick the right portfolio to suit their investment needs such that they 'confidently see' tomorrow's value today. In financial and investment services, therefore, it is no longer just about predictive modeling, but more about how you represent the models in a more physically intuitive fashion.

Need more? Remember the sensation that the movie 'Avatar' had created back in 2009? With visual effects to the tune and fancy that were never seen, felt, and in no way experienced before, it is not surprising that the movie has garnered $1 billion in just 19 days from its release, and $2.8 billions on the whole. The movie has opened our visual sensations to a dramatic range of colors, visual effects, life forms and nature such that we humans can never experience in reality. If we can create an altogether novel experience from a bit of technology, it should be our cue to apply this 'art' to our products as well. This will open a world of possibilities through our creative and innovative solutions that will become more than mere technology, more than mere solutions for a business urgency, need or pain point; solutions that will become art-forms, no less.

A Michelangelo of today can well be designated to work in something as distant from art, as the platform of trading and investment solutions with this altered approach.

Technological Trends That Are Redefining the Future

Technology changes lives; it can bring a sea-change to what we do, and the way we do what we do. In fact, it also does change everything about what we cannot do because what would have once been termed as science fiction is a reality or necessity of our daily life today. From the telephone to biometric access to your apartment, it holds true for all spheres. Once upon a time, dealing with a bank involved a physical visit to the branch. With ATMs (Automated Teller Machines) and online transactions, most of the transactions now happen without needing to physically interact with banks.

Disruptive technologies are slowly phasing out what seemed to be advanced technology about a decade ago. Even the ATMs are on the brink of extinction because more and more transactions are happening online, and most people do not carry hard currency anymore. While we know that things are set to keep changing with the passage of time, we do not really know which way things would swing. For instance, we do not know if cryptocurrencies would eventually replace paper currency.

In the recent years, the world has been particularly swift in the adoption of new ideas, trends and technologies. A few of them:

Cloud Storage:

Once upon a time, people stored data on punch cards, punch tapes, magnetic tapes, floppy disks, compact disks, flash memory and the like. Although we still use flash drives or compact disks, we can very well do away with them. Today, most of us only need a computer hard disk, and a cloud storage for extended or safe storage of our data. After we have moved onto cloud storage, as long as the servers at some place, where all our data is stored in a much-scattered manner, are able to serve our data security, integrity and availability needs, we do not worry.

This way, the cloud has become like a personal secretary. We just delegate the task of data storage to the cloud service provider and forget how it is managed. Today, IBM for instance, provides storage solutions to 49 of the 50 top global banks, and ensures that the data services do not get disrupted in any manner, whether it is with respect to speed of access, data integrity and accuracy, or most importantly, data security.

A sample of how cloud services can come to be useful for Wall Street related business:

> Access to our investment accounts as well as statements and reporting services could go onto the cloud. People would love to access

their accounts anytime, and anywhere provided the services made available on cloud are completely secure. Further, customized reporting services can become a sought-after value add. Businesses that look forward to breaking the traditional access barriers are certain to put their customers on cloud nine!

Big Data:

All a business firm ever needed a generation back was a hard-bound journal to record transactions. Today, it gets very sophisticated with the literally boundless expansion of businesses beyond geographical limits. We are, in short, in an age of 'Information Explosion', as it is called. Other than purely business related information, there is all kinds of data getting stored and updated onto the Internet and servers around the world, round the clock.

Several companies like Google, Amazon, Facebook, Apple and Microsoft have, over the span of our digital lives, collected and stored loads of information that talks everything of consumer preferences, buying patterns, interests and much, much more.

We leave a good trail through our phone calls, apps we install on devices like mobiles and tablets, credit card swipes, online fund transfers and net banking, phone banking, and the like.

The exception may be that some of what may be happening on the deep web may be untraceable for the most part–but that may become a topic for another sunrise.

So now that we have petabytes and petabytes of data, what is next? This is where big data solutions come into the picture. All the mountainous amounts of data can be easily captured, securely stored, analyzed and retrieved to provide meaningful sets of information to the user. The architecture and the algorithms that go into the maintenance and retrieval of data are varied, and serve all complex business purposes and ends by aiding the decision-making process in a way that no human being can. The field of data science and analytics is, therefore, one of the hotly sought-after disciplines in today's computing and business spheres. The purpose that is served: a store-house or rather, a mine-house of information that can be used to enable businesses as well as individuals better.

A sample of how big data technologies can come to be useful for Wall Street related business:

> Unless you end up as just another data hoarder, big data solutions can do a lot to turn your business around. Collecting and analyzing consumer patterns and preferences alone would enable you to design and customize portfolios that would suit their needs and

targets better, and bring positive returns on investments.

Cryptocurrencies:

Though the impact of cryptocurrencies on the global fiscal scene is yet to be seen, we have already jumped on a bandwagon and there is no going back.

The traditional or mainstream currencies were backed by a precious metal like gold, or as is the case now, by the 'guarantee' of the government of a particular nation, i.e., fiat currency. As an instance, the United States of America has abandoned the gold standard back in 1933, which means that the US currency (US Dollar) that we trade today is not backed by gold, but by the full faith and credit of the government of the United States.

Cryptocurrencies, unlike the run-of-the-mill mainstream currencies which are 'on paper', are electronic in nature. These currencies may be backed by anything from gold to quite literally, nothing at all! They are getting popular because transactions can be instantaneous, as the money can 'change hands' in a digital transaction where no banking or governmental regulations have any hold. Further, the transaction costs and fee involved in the use of digital currencies are significantly very low.

Cryptocurrencies that are backed by gold include ZenGold and OZcoinGold. PinkCoin, the first diamond backed cryptocurrency was recently on sale. The independently appraised stones are insured with Lloyds. The appreciation of this currency is based on the fact that the supply of colored diamonds is on the decline. (Australia's Argyle mine, which is the source of 90% of the world's pink diamonds is on the verge of drying up.) Bitcoin, on the other hand, is backed by nothing. Bitcoins can be 'mined' by solving mathematical puzzles, and the way the system is built, it gets complex as the time goes by. The level of difficulty determines the number of Bitcoins mined. There is a ceiling for the maximum number of Bitcoins that can be mined, though–a sum total of 21 million coins.

Every cryptocurrency, is again, inherently different in the purpose it serves or how it serves the purpose. Some of them are very significantly different from the more popular Bitcoin. Litecoin, for example, purports that transactions are processed at a much quicker rate than with Bitcoin. Unlike Bitcoin that requires special hardware to do the mining, Quark coin uses layers of the advanced encryption technologies such that any normal CPU can be used for mining. Ripple on the other hand, does not require mining at all. It is a currency (symbol: XRP) as well as a digital platform to transfer any currency other than just the Ripple in an instantaneous fashion, just like one sends information across the Internet. They call this 'Internet of Value'. Titcoin deals with an altogether different dimension of needs. It is setup for people wanting to access adult content in a secure and fast manner, and more importantly, it provides anonymity.

While the legality of the use and exchange of cryptocurrencies is yet to be established in several nations, many business concerns are already exchanging their goods and goodies for cryptocurrencies like the Bitcoin, Litecoin, Ehereum, Bytecoin, Monero and Zcash. The most popular of the cryptocurrencies is Bitcoin, which holds the largest market capitalization.

Users of digital currencies and cryptocurrencies, however, have to be wary as much of the investments in these currencies are speculative and tend to become a bubble. Also, the lack of a regulatory authority in itself has its own inherent risks to go with the benefits. Some governments are vehemently against these currencies for the scope they may provide for unregulated transactions and money laundering activities.

Some examples of the fiascos involving digital and cryptocurrencies:

- Pecunix became popular in the early 2000s as a gold-backed digital currency. The Pecunix system, which was the network for enabling transactions, was to be replaced by an open source, state-of-the-art system called Voucher-Safe System and PX-Gold. An announcement to this effect was made by the administrators, and the login for Pecunix's website was temporarily disabled. The voucher payment system did become operational, but PX-Gold never came back into existence, with the result that a lot of account holders lost their money.

- Liberty reserve was a digital currency service that held huge promise for investors back in the early 2000s. It was shut down based on federal investigations into fraud and money laundering, and overnight, the site that had a million users went bust. The founder, Arthur Budovsky, was sentenced to 20 years of imprisonment.

- Coinye, initially called Coinye West, became a defunct cryptocurrency because of a trademark infringement lawsuit for using the likeness of the popular American artist, Kanye West. This finally resulted in the developers of the project dumping all their coins and leaving the scene. The fate of all the investments made till that point of time is anyone's guess.

- More recently, 'The DAO' (There are several DAOs or Decentralized Autonomous Organizations with Bitcoin being one of them) was launched on the 30th of April, 2016, by a startup in Germany–Slock.it, only to become a target for hacking. It had spiraled down pretty much as fast as it shot up in value. It had created a sensation in the crowd funding world, with over $150 million raised from enthusiasts in the ICO (Initial Coin Offering) crowd sale. An unknown hacker made use of a known vulnerability in the system, and drained the 'ether' out before the fixes could have gone into the DAO. A 'child DAO' was created by the attacker to transmit the funds, and what made matters worse was that all the ether was parked in a single address, making the 'steal' an easy cake-walk. Unfortunately, the fix after the loss was not an easy one. People could not get enough consensus to quickly fork The DAO and keep the system from the risk of draining more funds into the child DAO. Although there are some work-arounds for a problem such as this, they beat the very point of having a decentralized, trustworthy ethereum blockchain. For example, 'buying in' 51% of the nodes to make a change and undo a transaction is a workable solution, but it breaks the very building blocks of a blockchain system's trustworthy and secure nature.

It is a point worth mentioning that 'Maker DAO', another system running on ethereum blockchain had also been identified with a similar bug as The DAO. However, since the system was in testing, the issue was quickly resolved and no harm was done. Keeping the ethereum safe for transacting requires testing and certifying the way smart contracts work.

A sample of how cryptocurrencies can come to be useful for Wall Street related business:

The craze for investing into non-traditional instruments is on the high, and you can use this to garner business by offering your clients a good portfolio that would aid them to invest some potential money into cryptocurrency asset classes.

Blockchain:

Blockchain is the disruptive technology that is set to reinvent a lot of processes in the banking, finance and investment domains alongside a lot of other areas of business and commerce. It has revolutionized the very basics of the financial industry by distributing and de-centralizing the ledger system. Unlike the conventional model where a ledger would be centrally held, blockchain ensures better security and integrity of information by doing the exact opposite–following a distributed system of ledger maintenance. Though its advent has been with the Bitcoin, it has gone way beyond just being a ledger for Bitcoin exchanges and transactions. Given the immense scope of implementation for this technology, several blockchains have come into existence, one of which is the Ethereum blockchain.

On a blockchain, transactions are recorded in the most secure, transparent and instantaneous fashion. Each new transaction would have to be approved by all the participants on the chain, which ensures that the transaction is validated. Each transaction is recorded on a block with the whole of the information on the previous block. A 'hash', which is a unique code, is generated from applying a cryptographic algorithm, and this is a tamper-proof way of recording transactions. This way, any given block on the chain can be traced back all the way to the originating block, and every block encrypts the transactions such that there can be no modifications or errors in the record once an entry is placed on the block. The data cannot be corrupted due to the factor of cryptography plus the validation needed from everyone on the network for any change that will need to go into the blockchain. Simply put, this is going to be a chain without a weak link.

Transactions becoming pretty much instantaneous means that we would be able to wipe out the need for any delay with the settlement of trades on Wall Street. If the industry is merely glad that we have moved onto T+2 settlement, it will be relieved to be on settlements within the minute–a haven for all parties concerned. This means that financial transactions, including cross-border and international money transfers, clearance & settlement of trades, proxy voting, and a whole lot of other business areas will undergo a massive overhaul, and with it all, the regulatory compliance requirements will be set to change as well–for the better. Every year, an estimated 40 million is spent globally by financial institutions just to fulfill regulatory requirements. With the implementation of blockchain, businesses will realize major costs savings, besides being able to transact with ease, accuracy and speed.

A sample of how blockchain can come to be useful for Wall Street related business:

> Blockchain is the new mantra on the list, and as stated, it can turn a lot of things into a different order from the way that they are today. Clearance & settlement services, back office operations, proxy voting–any area can be a potential one for building and implementing newer solutions. Although a few products may be largely dependent upon

regulation to come into existence, like in the clearance & settlement area, others may be more independent of it, like in the case of proxy voting.

According to the Spanish bank group Santander, banks alone can save up to $20 billion every year using the blockchain technology.

Internet of Things:

If the Internet connects systems and networks the world over, even bigger will be the potential size of the network when we talk about the IoT (Internet of Things). A research report from Gartner estimates that there will be 21 billion devices online by the year 2020. Cisco pegs it even higher, at 50 billion devices. On the other hand, a Morgan Stanley report predicts that the number of devices connected to the IoT will reach 75 billion by the year 2020. These estimates may even be conservative for all we know, because the developing world is gearing up at unprecedented speeds in adopting technology. Smart devices connected to the chain of the Internet will make it easier to communicate amongst them and interchange information. Also, it becomes easier for people to remotely control all their electronic gadgets, be it their personal computer, be it their microwave oven.

People may be able to 'cook' their dinner while on their drive back home from their work place. Talk of life made easy! Smart gadgets to smart homes to smart cities, even a smart world is not very far away. This is going to be a huge boost for businesses to sell smart products and smart solutions to the tech savvy generations of today and tomorrow.

Given the immense possibilities in this area, it is not surprising that the government of UK has allocated a total sum of £400,000,00 for research. Businesses involved in this area are attracting multi-billion dollar investments. While all of the reports by the pundits give vastly different figures, it suffices to say that no more needs to be spoken to indicate what is in the offing.

A sample of how the Internet of Things can come to be useful for Wall Street related business:

> Imagine the value it would add if your portfolio could be automatically managed to include asset holdings based on real-time situations in your life. A data feed from your car with regards to a bad crash, or data from your personal fitness monitor, could suggest a need for funds, and can make your portfolio have more liquid asset holdings until you tide over. After all, what is the benefit of holding

something of value when you cannot get money for an emergency?

Artificial Intelligence:

The human race has been enamored by robots for a long time back into its history. The field of artificial intelligence and machine learning has been gaining prominence over the ages, to create and make way for machines to do all things human, and in an exponentially fast and efficient manner. The first robot was the creation of George Devol, who had in the year 1954, made 'Unimate', which was sold to General Motors in the year 1961 to lift hot metal pieces from die-casting machines. From the popular kid's favorite 'Small Wonder' TV series, where a robotic engineer created a robotic daughter resembling a human in every aspect, to the movie 'Terminator', where a cyborg assassin comes after his human enemy, people seem to have received the concept of robots well enough. The reality of having intelligent robots within an affordable budget is getting closer by the day with more technical advancements being made in the field.

Companies in varied fields are looking at the potential cost savings that they would stand to gain if they invest into building programs and machines that can learn, improve and provide better value to their customers and clients, as compared to their human counterparts. According to reports from a Narrative Science survey conducted by the NBRI (National Business Research Institute), 62% of business enterprises would be using AI by 2018.

Indeed, the rise of the machines is going to become a reality in no time. Cognition and contextual understanding built into programs will leave a very thin line between machines and humans, if at all. Further, as humans cannot compete with machines in terms of speed and accuracy, they are left with the one choice–compete to build the best AI systems and technologies out there.

There are a several products that are helping businesses through AI. IBM's Watson is on this path– 'Watson Discovery' is a cognitive search and content analytics engine meant to aid decision making by being able to discover hidden patterns in the data that is ingested into the system–not what humans can do with ease.

A sample of how artificial intelligence can come to be useful for Wall Street related business:

> Machines can step in where humans cannot. The use of artificial intelligence in advisory solutions has already been showing huge gains. Although this may be one of the most

common areas for using AI in solutions, there is a wider field for application.

Artificial intelligence based analytics can build newer investment strategies that can give more promising returns as compared to traditionally followed strategies. It is like a disruption within a disruption. Feeds from social platforms across the world can make market trend and situational analysis a whole lot easier through AI. Managing back office operations or resolving reconciliation breaks can become totally simplified and error-free.

Crowdsourcing and Crowd Funding:

Innovations do not wait for time nor do they wait for money–anymore. When it comes to an idea, the time is always ripe for its implementation. In fact, with this fast-paced world, the time has already long past gone when it comes to bringing a new product to the market. This is one of the reasons for the immense success behind the idea of pooling-in resources from crowd sourcing and crowd funding projects. Today, the possibilities open to business houses through crowd sourcing and crowd funding projects is simply amazing, to say the least. All it takes is a notification on the Internet, and from a small idea in somebody's head to a finished product is a matter to be taken for granted–because the grants come in easy.

A stunning example of a crowd sourcing project is a virtual choir organized by Eric Whitacre, with 2052 voices singing 'Sleep', contributing from every part of the world. This man has choreographed the superb piece by being passionate about virtual reality and music, and getting them both together to create an experience that people would love not to forget. And this was just one of his compositions; there came more of them along the way.

A sample of how crowd funding can come to be useful for Wall Street related business:

> Equity funding is gaining a lot of momentum with the immense success of the crowd funding concept across the globe. Platforms that enable crowd equity are seeing a lot of 'footfalls', and startups are able to get their products faster into the market. Any project that shows promise, can take off in the most agile manner, because the wait for funds, and for the right hands is wiped out through a much-connected world.

Going back again, technology changes lives. Today, we should be more poised for change than ever before, given the dramatic benefits we can reap out of the seismic strides that technolo

How the Fourth Wave of Technology is Changing Businesses

Once upon a time, people just went to a brick and mortar store and bought what they needed. Life was pretty simple those days; it was also just plain boring. Today, people are not happy or satisfied producing and consuming the usual, run-of-the-mill things available in a traditional store anymore.

There is no wonder why we have people working on all kinds of permutations and combinations to get an edge over the competition. Every little extra bite into the market share gives a nice temporary glow of triumph, and then we are after more. This is the simple trigger for all the innovations that we see today, in products from food to clothing to machinery.

If you are one of those folks who use wallet services instead of paying by the traditional cash or card mechanism, you know how well people are adapting to newer ways of doing business. In this world, one thing links up with another, and we have an entirely changed routine of life after that. Back in the year 1979, Michael Aldrich invented online shopping by integrating an altered version of the television, the telephone, and a transaction processing computer terminal. When people first 'logged on' to the World Wide Web in 1991, people would have little realized that the phenomenon of online shopping would, one day, eat into the profits of plain brick and mortar only business houses. Now we have online stores that offer every conceivable product through a wide variety of payment options–you pay by cash, through your bank's online platform, through a debit or a credit card, a wallet service, or even through cryptocurrencies like the Bitcoin. Sellers are happy to ship to any corner of the world with the added benefit of not just product warranties, but guarantees of free returns and replacements. Even the delivery of products is undergoing a massive change, what with drones delivering packages.

To cut the long story short, the moral of the story is simple–invent innovative solutions that the customers can get with ease. It is only then that a business will be able to relate very well, both with the ever-changing trends in business, as well as the ever-changing customer needs.

Kings tend to be demanding, and the customer is our emperor. It means that we have to go out of our way to do what they may demand, and more. That is when profits are made and the shareholders are happy with their returns. To put it lyrically, this is the scenario when God is in his heaven and all is right with the world!

The fourth wave of technology–virtual, augmented and mixed reality, has literally transformed every conceivable domain. A few examples are given below to get but a slight taste of how transformation is becoming the new norm everywhere. As we see a few benefits that may be derived from this transformation for any one industry out of a whole lot, we would no doubt open our minds to take this business seriously. The end of the chapter contains further, less elaborative examples for the souls that would like to see more and perhaps, dig into further.

Virtual Reality:

Virtual reality is making the virtual world realer than the real. Today, the world is becoming more virtual as people have started to realize the convenience that it provides. Modeling, imaging, graphics, sound effects and the excellent mélange of arts with science is opening novel avenues for people to experience things differently, and add value to lives.

Virtual reality in the real-estate sector:

When one thinks of real-estate, one would definitely think of site-visits. Right from the planning stage to building the property and handing it over to the buyer, there are innumerable telephone calls, e-mails, personal meetings and visits to the site of the construction. As one can imagine, all of this would take a huge amount of resources in terms of time and money, besides the overwhelming physical and mental energies that would need to be pooled in.

In olden days, people went for real, mini size models of the architectural marvels they were intending to construct, although this was reserved for a few elite clients like barons or nawabs. As time advanced, table-top representations of sketches on chart papers became common. With the introduction of CAD (Computer Aided Design) software, people have switched over to on-computer models. This enabled one to not just see the design evolve on a computer screen, but also, do any of the following that was not possible before the CAD days of engineering-design:

- Design of robust 2D and 3D plans
- Simulation of shadows and making use of photovoltaic panels
- Filling the texture and color options for proposed building material
- Design of circular roof models, complete with intricate art
- Emulation of stair case designs, and much more, in a matter of minutes and without the cost of re-work for design changes.

Still for all, it is a picture that is just a few more steps closer to looking at the proposed structure, and it cannot become life-like in its appeal. Therefore, architectural designers and engineers are now modeling and viewing their engineered models virtually. Prospective investors can, by the virtue of virtual reality, literally get a feel akin to moving around in the property, room to room, and getting the most life-like experience made possible. Professionals of today can seamlessly convert a software generated 3D design or a BIM (Building Information Modeling) prototype into a virtually tourable model. The delight in owning a property comes alive for anyone who goes through a digital walkthrough of the virtual property, and much ahead in time before the property's construction is begun. Ceilings and flooring, walls and furniture, potted plants and swimming pools, including koi fish swimming in a room-length aquarium can come to life in the virtual world. It will help making all design elements get in sync with the plan, and also entice the prospective customer to seal the deal early-on. For people who can barely make a choice on which pair of pants to purchase, making a decision on buying a whole building can become a matter of a few minutes.

IrisVR is sought by professional architects, designers and engineers to convert their designs from software like SketchUp, Rhino3D, Revit and several others, to render a virtual model that the user can experience on devices like Oculus Rift and HTC Vive. All of this makes designing, building and delivering quality to your clients a better and easier experience. The most important aspect of all, is when you involve your clients in the design phase, as it puts the builder and the buyer in perfect sync. The benefit of it for a real estate man–your client can simply walk through his future home or office even before you have laid your hands on a single brick or building block. The client will know what he will be getting and you will know what exactly to deliver–a great boost for business.

As an example of how some of the latest headgears are changing our real environment to display a more enticing virtual one for us, let us look at Oculus Rift.

The Oculus Rift headgear comes with a stereoscopic display, a refresh rate of 90 HZ, high resolution of 2160X1200 and a 110-degree diagonal, and 90-degreee horizontal vision-field. With the position tracking system of 'Constellation', it efficiently tracks the head-position. Devices with the Oculus Rift and the Oculus touch controllers working in tandem provide the best rendering, by accurately tracking both the head and hand poses–all at a near-zero latency, and at a sub-millimeter level. This virtual reality, is also completely room-scale.

If this is not today's literal teleportation, then what is?

Augmented Reality:

In augmented reality, the elements of the real world are augmented. By the effective combination of the real with the virtual, the user's perception is enhanced several folds. The devices that you use become, in essence, your extended sensory organs. They will enable you to perceive a whole new and different world, a world that is an augmented version of the real one. One way to do this is through headgear, but there are alternative options. Instead of the bulkier headgear, one may even use lighter eye glasses. One may even carry a mobile phone with the AR solution in it.

Think of how 'Pokémon GO' created a new wave of AR sensation and went viral in the world of gamers. There are innumerable ways in which AR can be a great asset to humanity. Thinking of the possibilities it opens up in the medical field alone, lets us in on the fact that technology could add boundless value to life. In fact, it can breathe life into life.

Augmented reality in the healthcare sector:

In MIS (Minimally Invasive Surgery), doctors insert a camera into the body of the patient to perform a surgery. Take for instance, a surgery to the spine, which could be quite risky given the implications it can have. With the aid of AR technology, a 3D augmented image of the patient's anatomy can be viewed by the surgeon to perform the medical procedures accurately, much to the relief of a patient who may otherwise be terrorized by the very prospect of getting under the needle and the knife.

Philips is in the process of developing an AR surgical navigation system by which surgeries of a complex nature, like neurological or spinal, can become much simplified. The doctors are going to be better equipped to do their job of improving and saving lives.

Droiders, an official application developer for Google Glass, has launched 'MedicAR'. With the use of this platform, surgeons may view an augmented reality display of the patient on their screens. The procedure involves temporarily tattooing a target to the patient's skin, and directing the Google Glass to this target. The whole operational procedure would be right before them in a step-wise fashion. This is a wonderful tool to assist doctors, even with complicated operational procedures. It would also come to the aid of future doctors to be well trained for the profession.

Mixed Reality:

In the way that two brains are better than one, a mixed reality environment, where the real meets virtual, certainly adds more value to some applications. Otherwise called 'Hybrid Reality', mixed reality allows for the interaction of the virtual objects with real objects to create the desired environment, and that, in real-time. It has an immense depth and breadth, given the fact that there can be a huge number of possible permutations and combinations of the interaction between these real and virtual objects, and thereby, a huge to boundless flow of derived events. The 'Mixed Reality Continuum', as it is called, covers the entire range of possibilities between the completely real to the completely virtual.

Whether it is Elon Musk's SpaceX that is ambitious to achieve interplanetary missions, or the NASA, space science is looking forward to the possibilities that remain to be explored and realized using mixed reality. All major players in the market are vying for a share of the pie that mixed reality promises.

Microsoft HoloLens, for example, has a thousand ways to enthrall the users with its holographic technology that brings the digital to life.

Mixed reality in space science:

Training astronauts and launching them onto a mission in space is also a complex thing to do, not just launching a satellite or sending a rover into outer space to get back with information that humanity seeks, for both serving our inquisitiveness as well as our advancement as a civilization.

NASA had been in the news with its simulator for the ISS–International Space Station–a mixed reality simulator to aid astronauts in the making. In partnership with Unreal Gaming, NASA has taken training its astronauts to a whole new level. Trainees are required to experience, and most importantly, endure a zero-gravity environment and move about in a space shuttle to fulfill critical missions. The 'Neutral Buoyancy Lab', a physical facility with over six million gallons of water was the 'pool' for an astronaut in training. Trainees had also to attune themselves to 'factors' through a mockup facility called 'space vehicle mock-up facility'. This is a model of the space shuttle–'Orbiter', and it integrates the instruments and components of the International Space Station.

With the new application, all of that gets simplified. The 'Active response gravity offload system', as it is called, provides an immersive and 3D mixed reality experience that is close to life so training is a much-enhanced experience. Besides, NASA would be able to use it effectively for the design of habitats.

As anyone can see, this is sweeping technology. This massive technological tide is going to sweep over our world, and there just might be no area of our life that will be left untouched in the long run. Considering this, it is an easily conceivable fact that more and more applications will unfold before our human kind to transform our experience of everything.

They say that the sky is the limit. For those with the intent, the creativity and the capability, it should not become a tall order to put an escalator into heaven. What cannot be conceived in our imagination today will become something very usual in the day-to-day life of a future date. Why does one even need a time machine now that we can travel all spaces of time and geography, including making interplanetary visits through an innovative mixed reality video? If the question is just how far can we go, the answer will be that the lengths we can get to, will not have a measuring yard stick to state the distance. That is how far we can go with innovation!

Now for the souls who may want to look at more, there are immense ways to help yourself to the technical wave that is changing the face of businesses. Some of these product ideas may have already taken shape, while some may still be germinating. It does not matter, because in any case, we are all set to ride this wave:

- Our kids will no longer need tickets to go on an educational tour. They may engage the best virtual guide/avatar to get them on a tour to pretty much any place in the world. Children with special needs can benefit by attending the classroom right from their home or care center through the creative use of mixed reality.

- In the future, we may prefer to watch sports on our headsets better than we would want to watch them live or on a huge LED television screen. Virtual reality cameras like Nokia OZO, Samsung Gear 360, V.360 and several others enable recording in any direction or angle, and present a single visual, 360-degree field of view. Now one can watch an action-packed soccer match even while lying back on the couch, although the lying down part is more of an allusion; the soaring adrenalin levels would not let us do that. Better yet, we may also play on the field with the likes of Cristiano Ronaldo, Messi or Ronaldinho with the help of wearable gizmos.

- Mixed reality technology can be used to simulate the environment of an extra-terrestrial planet for scientific research and studies. Imagine mimicking the atmosphere of an alien planet and then setting up a human colony on it through augmented reality. It would become an interesting research and case study for anyone interested in colonizing the space.

- Ever wondered how wonderful it would be for us to create a movie with avatars of ourself with family and friends? Life's important moments and events can be better cherished when we can bring our digital selves to life through augmented reality software, enhanced with some fantastic audio-visual effects. It will definitively be more fun than an ordinary video.

- Imagine the day we would no longer need a mobile or a computer screen to google things out. Our very spectacles will detail us everything about the objects around us the moment we set our 'digital eyes' upon them—it would just be a matter of wearing a pair of smart glasses that would look no different from a pair of ordinary spectacles.

- How much would it mean for someone to be able to use immersive virtual reality and drown their physical pain or bodily challenges and troubles as their headset lets them drown in some deep-sea diving experience, or something which is pleasant to their senses? Imagine the sea-change it can bring to patients suffering from a post-traumatic stress disorder to be able to find healing through immersive reality.

- While on the one hand, the world is moving towards driver-less cars, on the other, one who would drive a car in the future would have a very effortless driving experience with augmented reality aiding him with navigation, street-signs, traffic and weather conditions, all displayed right in front of him on the windshield.

- Movie-making can become expensive, particularly if there are action sequences that get triggered with user intervention through augmented reality. However, it could become an interesting mix of movie cum gaming for the masses to enjoy, and an even better avenue for producers to garner revenues.

- The value is much augmented when our devices get connected to the Internet of Things (IoT). A production floor supervisor, for example, would be able to take stock of machinery on a production line just by walking past each machine and gauging essentials like the temperature, RPM, oil levels, or whatever it is that needs to be measured to ensure things are working fine.

- Oil rig or mining simulation can help companies not just train personnel, but also evaluate the effect that weather conditions can have on their operational expenses. Improving personnel's safety and increasing operational efficiency can have a positive impact on the overall performance and profitability by leaps.

- Haptic technology can be used to implement training solutions for drivers to prevent the mishaps resulting out of negligent and reckless driving. To simulate how it feels to be in a bad car crash can deter the driver from ignoring the right track, or the right way to drive. It is more powerful than road-signs and slogans to remind and to tell a driver that mistakes, even if they are small ones, can be very costly.

- Using haptic technology, people can 'connect' with others in the more literal sense of the word. Instead of an audio or a video meeting, people can have mixed reality meetings with anyone from any part of the globe. Mixed reality solutions can help confer with holograms of other people, and even facilitate hand-shakes that would seem to be real.

- Simulations of different terrains and weather conditions, elevations and atmospheric pressures, visibility levels, and besides these, combat sequences including guerrilla warfare, can aid military training without the cumbersomeness of getting everything right in a completely real and physical setup.

- Even the cosmetic industry is going to be in for a giant leap of change. Using virtual or augmented reality, make up artisans and professionals, as well as non-professional individuals wanting to try out the look of the season, or experiment with shades, styles and colors, can just sync up their photo with a number of makeup options and choices on a simple mobile app, and apply only what suits and would turn heads around. A similar solution could be applied to choosing anything from clothing to accessories like handbags, shoes or sunglasses.

- For architecture lovers, engineers can blend different styles to create their dream palace, bungalow or house. A mishmash of Georgian, Renaissance, Gothic, Indian and Chinese architecture can become reality right from one's creative aspirations. Out-of-the-world expectations in design need not be expensive or disastrous anymore, when there is technology to assist with making it as good as it can get.

The list can go on further, but this is enough sample to see the impact the fourth wave can have on our lives.

Investments and Wealth Management at the Cusp of Change

People, companies, corporations, and governments alike, make investments for a variety of reasons. There are innumerable tools and platforms that are needed to meet the needs of the investment and wealth management chain. There is a whole gamut of solutions to meet the varied needs at all the different links in the chain, and this includes the following:

- Investment account and portfolio management
- Tax saving
- Cash flow management
- Investment and advisory solutions for different instruments like stocks, fixed income, options and mutual funds
- Family business and financial advisory
- Retirement
- Goal based investing, like in the 529 plan
- Wealth management and estate planning
- Investor education
- Trust management
- Tax management and tax loss harvesting
- Regulatory compliance
- Philanthropic planning
- Custodial and trust services
- Foreign investments
- Corporate funding

Each of the above areas involves seasoned professionals, tools and technical aids to meet the stated goal. This would imply that there is a need for a multitude of solutions, and solutions that will need to keep scaling up to meet the ever-increasing demands in the market and the technological advancements of today. This makes the field of investments flooded with multinationals and specialized companies offering everything from retirement solutions like the 401K and 403B plans to the regular tax paying individual, to providing the best of investment advisory solutions and portfolio management for the institutional giants as well as for people with huge fortunes to manage.

A variety of aspects get factored into influencing and customizing one's investment options and choosing the right kind of solutions. Some of these are listed below:

1) The target client-base:

   The dynamics change with each bracket–individuals vs. institutional investors and investors bucketed into segments like HNV (High Net Worth), VHNW (Very High Net Worth) and UHNW (Ultra High Net Worth).

   For instance, there is a sea of difference between a product aimed at financial advice alone vs. one aimed at wealth management, for the former serves individuals with growing their finances and the latter caters to very high or high net worth and ultra high net worth clients to manage their wealth.

2) Goal of the investment:

- Tax savings: Tax saving through IRAs, 401(k), 403(b) or 457 as may be applicable for each individual case.
- Building savings: Building savings for the future, or in case a person has maxed out on their retirement plan. It could also be set for the sake of hedging against inflation.
- Goal-based savings: Saving towards a specific goal such as being able to afford a much-coveted holiday, purchasing a home, or meeting the needs of children's education.
- Growing wealth: Investing to ensure that the wealth a person possesses can be grown and multiplied.
- Emergency situations: Saving for unanticipated emergencies in the future.

3) Time horizon:

   The term or duration for which the investment is made; based on the goals, investments may be made for a short, medium or long term.

4) Risk appetite:

   Risk tolerance or appetite also makes for an important factor in determining the type of instruments in which to invest, and forms the core component for building a portfolio.

5) Age of the investor:

   The age of the investor has a bearing on the amount he can possibly set as the goal, the maturity period or investment horizon, risk factor, etc.

6) Market conditions:

   Although this is a very volatile factor, it still does set the direction to planning and growing funds.

7) Investment instruments:

   Towards getting that right mix for the portfolio, and because of investor preferences, this gets into the formula. The other factors influence the exact percentage-allocation of the instruments in the portfolio mix.

Irrespective of all of these aspects, the target of any service is to finally make money work for us, and not the other way around. Firms that have found the best way of doing this are always on the winning line, and beat all odds of the market. With the times changing rapidly in terms of the political and economic aspects of the world, with the way we are 'networked' today, an aspect that impacts the economic and financial conditions in any point of the world would soon have an impact on the rest of it. The only question is, how impactful exactly will the ripple effect be. It obviously implies that it is tough for anyone to be a leader and the service provider of choice on a constant basis. Anyone from an individual investor to an investment bank will have to be careful about the path they trudge while they invest.

Our efficiency in handling the business has a direct impact over the assets we hold. The volume of assets under management, for any given business concern is certainly of business concern because it will impact profitability figures in a significant manner. A lot of analysis, algorithms and models have to go into making the solutions to invest and grow wealth. As anyone in the business knows, this is by no means simple and easy.

Our traditional processes and tools cannot gear up to what the modern-day technologies will enable us to do.

There are all kind of players in the field. Speaking only of the wealth management space, there are independent wealth management advisors to mammoth, global investment banks playing for the pie with varying modus operandi. While the investment banks may only seek the ultra high net worth client segment, the independent wealth advisors may serve only a handful of clients on a commission basis. With the steady spurt in the number of high and ultra high net worth clients, the demand for better wealth management solutions is also increasing at a good pace. Today, it is all about specialization; therefore, the family offices are slowly shutting down and getting obsolete with the heat of competition from the specialized investment banks and wealth management companies like UBS and Goldman Sachs.

This playing field is also fast changing due to the demands of the current age, and even the big players are sweating to catch up with the trend. The disruptors are drastically changing the way the game is played. Today, people don't mind paying a little more for professional solutions that add value to their portfolio returns or serve their interests in a unique fashion.

Every business has, therefore, made innovation its key driver in business, figuratively speaking. The very same solutions are, therefore, getting re-invented to enhance efficiency of operations, predictive analysis, modeling, etc. all of the time.

Besides, there are strong reasons why we need to see our businesses at the cusp of change and this includes several factors that the next chapter would elaborate upon. But before we get there, I have included below, a few areas where we can ponder upon adding value through the fourth wave and serve our customers better. Ideas around these would follow a while later.

Automatic portfolio rebalancing:

The asset mix is obviously the magic mantra to investment gains in a portfolio. This is, by far, the only way to really manage the market's volatility. A rebalanced portfolio is better equipped to grow in the current market conditions and stay fixated to fetch the financial goal set at the outset. To do this, the asset manager or the portfolio manager would have to keep a strict tab on the rates of returns for each asset class in the portfolio holding, and rebalance the proportion of investments in the mix. However, there will be a financial cost attached to do this. It depends on whether the solution's strategy is a passive type or an active type, and whether the asset mix is more towards the conservative side or towards the aggressive side.

Typically, rebalancing is an integral part of investment and wealth management solutions, but there are factors that decide whether it is going to be one of the core features in the product.

There are several strategies for portfolio rebalancing, and which of those to consider is purely dependent upon the type of investor, the time horizon, the target financial goal, the asset mix at the outset and the amount of holdings in the portfolio. The dynamics vary largely, depending upon the type of investor– individual vs. institutional as well as the preferences of the investor. This is the reason why different strategies of portfolio management may be implemented for even those portfolios with very identical allocation of assets.

It becomes imperative, therefore, for the asset manager to choose the appropriate model for simulation, given the fact that there can be no single, optimal asset allocation that goes for all kinds of investors and investments. In this world, there is never that one size fits all strategy. The perfect portfolio is a myth, just as a perfect market is. With the complexity that is involved in getting the right mix of assets and holdings, and continuing to keep it right, one may offer just the threadbare solutions as against the value-added solutions provided by thorough-bred and seasoned professionals. There are any varying number of permutations and combinations as go-between options. While the basic solution may offer a good portfolio for meeting the target goal of an investor, a really good, top of the range product would have to provide automatic rebalancing to not just meet the target goal, but to exceed it as well as it could.

Customized statements and reports:

While all investment products will have to offer statements and reports of the performance of the customer's assets, there are innovative means to make these more useful for the investor as he glances or studies through them in the way that he would prefer.

Presentation of a statement should ideally depend upon the type of the investor and his preferences. While there is a little to the presentation aesthetics, there is more to what exactly is the point of the report, or the highlights for the customer to focus upon. While some may be interested to know just their net profit or loss, others may want a detailed report on portfolio adjustments made and proceeds, summarized on a periodical basis. Finally, statements, like ice creams, can come in different flavors. If one wants to be a forerunner in the business, they better do.

Impactful investments:

If we assume that the only goal of any investment and wealth management solution is to make more and grow more money, we would be wrong. Today, more than ever, people's attitude is undergoing a major change. The demand for 'Impact Investing' is seen to be on the rise. This is pushing companies to evolve their products such that the investment strategies they choose support environmental and social changes towards the positive. In other words, a firm that stands for a strong environmental cause would get more funding into its coffers. There is an increased demand for transparency into the way the invested funds are spent as well. In fact, as per a survey by Nielsen, 55% of respondents said that they are willing to pay more for products and services from companies that are inclined and committed to making a positive change in the society and the environment.

Today, some of the top players in the market providing investment and wealth management solutions are: JPMorgan Chase & Co., Morgan Stanley, Goldman Sachs, UBS, Barclays and Bank of America Merrill Lynch to name a few. The disruptors are soon catching up to have their share in the pie.

To serve as empirical evidence for the potential of building better products through the fourth wave technologies, I have provided a couple of examples below. But more importantly, their purpose is to remind us that a radical change in the way we build our products is only eventual. There are more that the reader will see in a forthcoming chapter, but this is to set the base for us to be receptive of ideas that call for building and implementing innovative solutions.

1) Comrach's Case: Comrach has launched a VR prototype for managing wealth. The app enables investors to get their investment strategy in sync with their objectives and needs. The aim is to bring about disruption through all of 3D goggles, algorithms, human relations and fun put together. An interesting way to invest and to build wealth, indeed.

2) 8ninths and Citi innovation lab teamed together to come up with holographic workstations that empower traders through Microsoft Hololens. With this, Citibank traders can now work more efficiently with hundreds of financial instruments by volumetric data visualization though 2D and 3D elements in MR. This means that traders can, at just a glance of the representation, get an accurate picture of the trading environment

to do their jobs way better than it was possible before.

As stated earlier, we are at the cusp of some serious change. The dawn of the age of millennials and centennials poses several new challenges, as does new opportunities. Besides, there are some more factors to add-in as the drivers for change. These, we will now see…

The Age of Millennials and Centennials

The term 'generation gap' holds much significance because each generation of people, those born in the same 'birth cohort', have certain distinctive characteristics and values that more or less apply to all individuals in the cohort (with some variations across some demographic factors). However, with the world getting to be more connected today through technological advancement, people being able to communicate better than ever before through the Internet or mobile communications, the variance between some demographic factors is getting blurred enough. This means that more people of the current generations the world over, do exhibit more similarities than the previous generations did.

A reference table to see how the birth cohorts categorize the generations of our interest here:

Baby Boomers	1946 to 1964
Gen X	1965 to 1979
Gen Y	1980 to 2000
Gen Z	2001 to 2010

** On a side note, the baby boomers have been termed as such because they were born in an era of population explosion.

** The range of years in the table are not 'standardized'. The census bureau only classifies the baby boomer generation in the range of years stated above.

The Gen Y (AKA Millennials) and Gen Z (AKA Centennials and iGen) populations are increasingly different from the previous generations–the Gen X, and the baby boomers in their thought, approach and preferences on pretty much everything and anything. The primary contributing factors for this–technology, economics, and of course, parenting.

If one has to ask why we should be going around these facts and factors about the millennials or the centennials, we come to the crux–which is, millennials form the largest percentage in workforce, meaning our businesses revolve around their spending or investment choices, in the literal sense. The centennials are not far behind; we will soon have this population affecting our businesses and their figures in terms of returns and profits. With more wealth at their disposal, either through their earnings or inheritance, or both, the generations of millennials and centennials would dictate the rapid changes that have to go into our product offerings for savings and investments. Business gets challenging owing to the fact that these generations are increasingly demanding, are more tech savvy, and much more open to change. It is like a double-edged sword for business; it might kill you or make you an emperor in the bargain.

The traditional models in doing business are, therefore, set to change because of the influence of the younger generations and several other factors beyond just the generation factor. Below is an elucidation to figure out how the scene is set for change in the business of investing. More importantly, it also enables us to identify these as our levers for building our businesses. These factors present market dynamics requiring the very building blocks of our traditional businesses to be revamped. A disinclination or refusal towards changing the way we run will only get us to a halt. An inclination and interest to pour our research and development funds into this, on the other hand, could catapult us into the golden, booming era for our business in the future.

- About 50% of working population in the United States of America alone, are living paycheck to paycheck. This is a nice catchment area if only we could succeed in making this population realize the importance of saving and growing money. In short, we need people to become investment-wise.

- There is a kick-start problem with investing–many millennials look upon investing as akin to gambling. They may want to gamble in real, for fun, but with investments, most people are very cautious if not entirely risk-averse. Close to 80% of the millennials are not investing their money in the market, although studies show that many of them do save a percentage of their income. These savings, however, are not finding their way into the world of investments. Most millennials work for the money, but do not think of making that money work for them.

- There is immense potential for business from the underbanked and unbanked populations. As per statistics from the 2015 FDIC National survey of unbanked and underbanked households, in the US alone, 7% households were unbanked. That is a staggering nine million households.

- Gone are the brick and mortar days and so are the stores and brands that have restricted themselves to doing business through physical channels alone. The number of dead-malls in the United States alone shows an alarming picture of the changing façade. This is a pointer to the fact that the business of investments also needs to move with the tide, and the tide is high with respect to being innovative. With the millennials hooked onto their smart phones and tablets, solutions have also got to be presented on these new platforms and have to be lean, and at the same time, feature rich. With innumerable channels for accessing the digital space at their disposal, the ball game has a newer and much wider screen today when it comes to selling to this generation; just forget the geographical barriers.

- The millennials are technology savvy to the extent that they talk more to their devices than to humans around them. A faceless person on a social platform like Facebook is not a stranger. People have learnt to build social circles online, without any physical interaction. People are ready to spare any expense to be immersed in virtual realism and change the world around them. This explains why VR headsets are becoming more popular by the day. With even book lovers taking to digital books through devices like the Kindle, and newspapers available online today, it is clear that the millennials are keeping away from anything to do with papers if there is a more convenient approach. There is no wonder that digital solutions are at an all-time high demand.

- The number of women in the work force is certainly on the rise. This opens a huge potential to capture business from female investors, where the sphere of investments was largely restricted to the male population just a few years back.

- Statistics indicate that millennials have more wealth at their disposal than the Gen X had, before them. They are also set to inherit from their ancestors, and with all the money, there should be money management to go hand in hand with, obviously. It is worth mentioning that there is some trillions worth of business to be tapped. The Wall Street, in the foreseeable future, is set to get exponentially busy.

- Increasingly more people are preferring offshore investments, meaning that the market is as big as the whole of the globe, unrestricted by geographical boundaries. Also, emerging-market investments are showing a lot of promise in terms of growth. As per the estimates of the International Monetary Fund, these markets are expected to grow two, or even three times the rate of the growth the developed markets have been seeing.

- Millennials are more interested in investing into companies that are more innovative, technically more sophisticated in their product offerings, or into companies that make a positive change to the world or the environment. This means that the portfolio mix will have to bear a massive shift.

- There is a marked increase in demand for Sharia compliant businesses, and not surprisingly, Sharia compliant investment vehicles as well. As a matter of fact, an Emirates based firm has recently unveiled a Sharia compliant currency akin to the Bitcoin in the cryptocurrency world. One difference is that this one is a gold backed cryptocurrency. OneGram, as it is called, seems to have a huge potential, with the followers of Islam being around 1.6 billion strong. An OGC token could be obtained by registering for a GoldGuard account and purchasing the yellow metal at live spot-rate.

- Millennials do not believe in the concept of brand loyalty, although they love brands. An increasing number of customers, therefore, are not likely to stick unless they consistently see value and increasing gains in a business proposition–all the more reason why getting them into your customer base, and keeping them there will be more challenging.

- An increasing number of investors are actively seeking higher engagement levels with their wealth managers to customize their solutions. They are more empowered considering that they have a wider choice of products and bespoke services on the market, to choose from. This makes tailoring products to deliver the desired value all the more tough.

- Millennials expect services of portfolio rebalancing and investment advice as a needed basic service rather than an additionally paid value-add. They expect automatic, and more frequent rebalancing of their portfolios to tune-in with the market, rather than revise at regular time intervals. Besides, this is a more sensible option given that the markets have become highly volatile.

- Millennials and centennials, having grown up in the influence of an increasingly digital world, certainly expect a lot more out of their portfolio statements. They would prefer to be more in control, than just look for a statement that comes in say, once in a quarter. They would be wanting something more decent, like real-time, to the minute updates! Most expect that they should be able to access their accounts on any electronic device online, and without a beating to their user experience. This means products have not just got to be available on a host of devices like mobiles, tablets and laptops, but also be compatible with a host of different operating systems.

- With increasing double-income households and increasing wealth in the middle class, some people might be able to invest almost half of the family income into financial instruments. Most people today have become more focused on saving for the future.

- Millennials give more importance to customized and personalized services. That way, they feel more valued doing business with you. It is not the least surprising, therefore, that many shops are now providing a lot of customization options, from coffee mugs to cars.

- With the advent of big data, capturing, storing and retrieving consumer taste and behavior has become phenomenally easy, and businesses can learn easily now than before, how exactly they can gain the fancy of a prospective customer while retaining their existing customer base. However, businesses can enhance the capabilities of catering to their customers better if they can combine their big data analytics with the analytics of thick data.

- Millennials and centennials have lesser time at their disposal, hence lesser patience. They would, therefore, spend more money on solutions that give them the freedom to choose the best solution for their needs in the least time consuming fashion. Most of them prefer to deal with businesses that do not require their physical presence at a location, or to expend time in traveling to offices and spend time filling out paper forms to process.

- With more people savvy about being entrepreneurial, and taking the risks and hardships that come along with their startup businesses rather than being a part of somebody else's grand design and staying at a salaried job, it is clear that millennials want more in return for their buck. They are definitely prepared to go a long way to get it all. This means that they would likely be expecting a great deal out of any business proposition. These generations are increasingly penny-wise and pound-wise too.

- These generations like things at a quicker pace. One look at everything from how the pace of our movies to the pace of books to the performance and speeds of automobiles and gadgets has drastically been on the rise would prove this. A quick solution for them, might be a good solution!

- The baby boomers simply loved the stock market. Perhaps it has a little something to do with gambling spirit, but a quick look at the yester years would tell anyone that investing was not seemingly as risky as it is today. Today's dynamics have possibly over-cautioned the millennials from buying those chips on the secondary market. The economic and stock bubble bursts of 2000 and 2008 are playing the devil's advocate on their minds. The need of the hour is a little trust factor, given the number of scams and schemes that have resulted in losses for millions of investors. This is precisely why the number of regulations is increasing, and why people may be wary to part with their money, because there are no guarantees with the business of investing.

- Millennials are comparatively more investment wise, no matter where it is they are putting in their money. They go by reviews, and make cautious and well-informed decisions when they spend, right from buying something as simple as a fiction-novel. If one would take a glance at how sellers on online stores are craving to get reviews and flaunting them off, one would realize that millennials are a race of beings that may spend exuberantly, but not without extreme caution.

Factor in all of the above, and it does not take a genius to conclude that there is yet untapped business potential, although it is more of a tall task to meet the expectations of the scenario today, and of the coming times. However, if the ancestors were not afraid to go gold prospecting in a wild and untamed land, ready to face and surmount both natural and man-made challenges apart from possibly taming their own mustang for a mount, the task on our hands should not be much difficult, though of course this plane is altogether different.

It is not an exaggeration to state that we have a huge pool of the potentially lost potential investors unless we create newer ways to provide the same investment solutions, and also introduce more innovative products as we go. Therefore, the solutions I present in this book in a forth-coming chapter would tie back in here, because they map to several of the aforesaid factors that need factoring in as we design our products.

All in all, it is challenging for the investment solution providers to tread the path. This is more so when you consider that the investment patterns are drastically changing, and it will only continue to be so as we keep going.

Products That Have Scaled Up the Innovation Ladder

In fin-tech, regulatory and legal requirements form a major road block for the innovators to surmount. However, as they say, where there is a will, there is a way. In fact, where there is an idea, it makes for a way.

As per a KPMG report, the challengers, as in the new breed of banking firms that are digital savvy, have reported an increase in total profits by £194 million, as against a drop of £5.6 billion for the top five banks in the UK–HSBC, Barclays, Lloyds, RBC and Santander, UK (not necessarily in that order). Traditional banks are rapidly losing out to wallet and mobile banking services for the convenience factor alone. Besides, novel products on the market are able to offer several other additional features or services that may not be available on conventional solutions. This is certainly increasing the pressure for the traditional solution providers to reinvent the wheel.

Any product offering that has to live in the changing times will have to keep scaling up on a continuous basis. For example, the advent of big data technologies is assisting businesses to see more, and do more to garner profits. But there is more to technology that remains to be utilized. We do not have to wait until a distant future to see the rapidity of change in product and service offerings in the industry. As newer ways of doing business are being realized, we see that fin-tech is actually a pace behind the rest of the industry.

However, it is catching on and gearing up, if only one gear up at a time. Several products in fin-tech have been invented, or re-invented to meet the demands of today and to be able to continue meeting the demands of tomorrow.

Just a sneak peek is provided here to see how disruptive technologies are forcing businesses to change the way they have operated for decades, and making them reinvent their products. Also, this gives sufficient proof as to how creative and innovative companies are actually dictating disruptions to create an advantageous position for themselves in the market.

Proxy voting through blockchain:

Proxy voting is the casting of a ballot by a shareholder through electronic or postal means. During the digital age, postal proxy voting has been phased out in favor of electronic voting, mostly held on a web-page hosted for the purpose. The advent of blockchain has provided a completely trust-worthy and transparent way to do this.

Broadridge Financial Solutions (NYSE–BR), a pioneer in providing investment solutions and investor services, has successfully executed a pilot of proxy voting for an annual general shareholder's meeting, in conjunction with JP Morgan, Northern Trust and Banco Santander. The company has relied on a private

Ethereum blockchain as a backup to the traditional voting software at the meeting of Santander Investments, and proved how the latest technology could be used to provide traditional solutions in a more reliable, and at the same time, transparent manner.

Investing through a personalized portfolio on a mobile application:

For a person without a clue about investments, Collective Returns, Inc. has come up with an interesting and innovative solution. Stash, as it is called, is a mobile application that enables the clueless to tread the path of investing with ease. Investors can buy fractional shares, choose a portfolio that meets their personality and matches with their interests, and keep an eye on the milestones in just a few clicks. A list of sample holdings for each of the profile types of investors, will certainly help a novice investor understand where and how his fund will be allocated in a very simple manner. Stash is thus, changing the world for the small investor who can now realize his dreams without getting bogged down by the chaos he would likely encounter in the investing terminology. All it takes is $5 to begin investing, as well as to learn about investing on the fly.

Clearing and settlement banking on the cloud:

Clear.Bank of UK is set to revolutionize the way clearing banks work. Without any dependencies on other banks in terms of operations, Clear.Bank, that is being launched in the third quarter of 2017, is set to transform banking by completely letting go of legacy architecture to provide a digital clearance platform on the cloud.

Advisory solutions through robots:

Vanguard's Personal Advisor Services is an expert advisor minus the emotional quotient of a human being. The service is supplemented by human advisors, but the core of it is the digitized aspect. It benefits investors by creating a tailored portfolio to meet the investor's goal in a simplified fashion, and it also rebalances the portfolio allocation automatically. It is no surprise that the two-year-old product has already got over $65 billion assets under management.

Mobile only banking:

N26 is a mobile bank that enables customers to open a bank account in under eight minutes, carry out instant transactions, transfer money to 19 different currencies, and do much more. This German bank is also soon to launch their investment platform where investors will be able to put their money into their portfolios from the convenience of their mobile application.

Innovative banking through community:

Fidor Bank, (acquired by France's Groupe BPCE) a German based digital-only banking firm, is challenging the traditional stereotypes in banking. Having launched its operations in the UK back in September 2015, this is one of the first banks that has started using Ripple for payments and transactions. Unlike traditional banks, Fidor enables its customers to participate in the bank's decision-making process. It is the recipient of several international awards for its disruptive innovation and transparency. It interacts with its customers on social media platforms like Facebook, LinkedIn and Twitter, and encourages community discussions on any aspect of finance. It has around 100,000 clients and more than 300,000 members in its community.

Income adapted loan payments:

Align Income Share Funding, that was started as Cumulus Funding, enables customers to take loans with an interest rate that 'aligns' with their income. Unlike in any traditional personal loan plan, the payments are pegged to the borrower's income, which makes it highly flexible, hence attractive to customers. The loaned amount also reaches the debtor way faster as compared to the traditional means, needing only one to three days of processing. These ISAs (Income Share Agreements) are aiding borrowers to have peace of mind because of the financial security they offer, with payments going all the way down to zero in situations where a person is at a loss of job, or on a pay cut.

The moral of the story–times have changed, and so have consumer patterns. It goes to imply that products need to change as well. Whether it is an investment banker, or a full-service brokerage house, a family wealth advisory firm with very limited clientele, or a product-based company offering solutions to these parties, today's primary need is to innovate–or evaporate!

Innovative Products Through the Fourth Wave–A World of Opportunities

As we have seen, people of today are not drawn to brick and mortar firms no matter how fancy they may be made to look. People know where to find value, and more often, it is on the digital rather than the physical plane.

Companies may perch themselves up on top of the big data mountain by collecting heaps of data on consumer choices and market trends, demographic information, and all of the relevant market intelligence. This may help them identify customer behavioral patterns, get competitor analysis, and make predictive analysis in several angles, but it is not going to help them if they only augment their business processes and products in a step-wise approach. As a point of note, many companies are only hoarding immense data which will only end up as 'dark data' that will not serve their business purpose. Even though a business might go the whole length of the data analysis chain, the next seemingly logical step, in most cases, may be unhelpful. A significant digression needs to be made and some very unconventional thinking needs to be applied to meet the massive shift in the advancement of technology.

Blockchain, Big Data, Virtual Reality, Artificial Intelligence, Internet of Things and the lot, are not mere buzzwords. It may still be a little too early to reap benefits, but not too early to lose out to competition. While some firms may be grappling with the mundane run of business affairs, others have already begun to implement technologies to benefit their business. Some companies still, are sandbagging–as in, investing into disruptive technologies (albeit some of which may still be in the nascent stages), secretively designing products that will put them at a market-edge and these are not going to figure easily in the run-of-the-mill market intelligence reports we may garner today. These companies will clearly become the front-line winners, those that will dictate the way business gets done tomorrow.

Speculations suggest that Apple is investing pretty heavily into augmented and virtual reality. Pundits are keeping an eye on trying to know the secretive development of Apple's projects in the field, and the fact that it has taken over companies like Metaio and Faceshift, is keeping everyone at the edge of their seats as they are trying to guess their game. (Metaio was an AR company based in Germany. It was acquired by Apple for an undisclosed amount on May28th, 2015. Faceshift was a motion capture and animation technology company based in Switzerland.)

Virtual reality and the related technologies will be huge, rather momentous game changers. For the complexity and intangibility-riddled industry of investment and wealth solutions, these could be rather a boon. Merely trying to forge into newer markets, pushing the geographical boundaries out, upgrading to meet regulation, trying to address the business problem better–these are not going to get us tremendous results; only investing our minds and money into tremendous technologies will.

When business is all about the customer's experience of your product or service offering, what better way is there than show them what they get by means of virtual reality? It is not a fallacy to state that perception is everything. Impressions we gain from our sensory perceptions are what make us invest into experiencing a service or product. Many people buy products that they have already experienced and been satisfied with. When buying a product or service not experienced before, they go by word of mouth and reviews, besides their own anticipated value quotient. In any case, fail in the perception of the customer and he is someone else's by the next moment. For those still in doubt, here is an elementary example–the icing on the cake, and the garnishing of a dish appeal to the customer more and get it on to the platter more easily than an unpretentious and simple looking dish. Of course, it has to appeal to the taste buds to have repeat orders– that goes without saying.

While it is easier to conceive the use of technologies like blockchain, cloud services, artificial intelligence and data analytics in the field of financial investments and wealth management, it is rather difficult to conceive what can be derived out of virtual reality and related technologies, unless one is very creative. This is something that involves both art and science to deliver what it can, and does. The fourth wave technologies, will be perfect to give appeal and tangibility to the otherwise less tangible products in the world of finance. An added advantage could be a bit of fun and entertainment by the side.

The following is food for thought into how we can innovatively implement the fourth wave of technology to benefit the stake holders in the business of investments and wealth management. The scope is infallibly immense. Research conducted at Stanford University's Virtual Human Interaction Lab has shown that the change in consumer behavior is more when we make the cause and effect relationship perceptual rather than theoretical. There is no wonder why show and tell kind of propaganda gains over mere statements from a company on how effective their products are, or media reports talking tall things about the company's service offerings. If it is not through one of VR, AR or MR, what better solution do we have for show and tell? Because seeing, is believing! If there is one thing that can propel anything, it is the human mind and the human mind is lead to believe what it sees, all else aside. An image is the biggest evidence that there can ever be, and that is why they say that a picture is worth a thousand words.

Before I proceed further, I would like to say that there is nothing outlandish in this world. The human tendency to strike off a thought or an idea as just fanciful, silly, or impractical is just that–a tendency. Until they have materialized, the telephone, the mobile phone, the Internet, wireless technology, nano-technology, 3D imaging and videos, X-rays, space travel, and products related to these fields have all seemed pretty outlandish.

Now some grub for thought on product development use cases using the fourth wave of technology:

Here are some use cases I present that could become business-drivers. These are but a few ideas, and in each, a lot depends upon how any business would like to expand the idea, based on the business firm and its organizational goals, product line, funding, markets it serves and the like.

Some of the ideas below revolve around gaming. The reason behind this is that gaming appeals to Gen Y and Z, your target customer segment the most. Also, most of these revolve around how to empower your end customer. This is so, because in business, the goal should not be left to empowering your own systems and processes, products and solutions. An intelligent business needs to empower its customers so that they can, in turn, empower the business by dealing with it on a time and again basis.

1) A VR game for investor education and mimicking investments:

As we have already seen, statistics show a spike in the use of mobile solutions and applications. At the same time, we know that most millennials are not very inclined to invest. A pretty solution would be to put the two together for enticing individuals to invest through a virtual reality mobile application. This could enable investors in becoming investment-aware, help them mimic real-world investments through virtual investments, and also choose to get advisory services, all in one shot. A few points on what this kind of a solution could feature and enable, in a sample case:

- The investor would create a fancy avatar of himself, which could be anything based on the theme that you select for the virtual reality app. It can range from a plain individual to an outlandish avatar, as in, an alien or a super hero. You can let your ideas take wings based on how you would want your application to look and feel, and the concept you choose for the game.

- He could walk around on the Wall Street, the Bay Street, the Dalal Street, or whatever place, based on the geographical markets you serve as an investment broker or product designer for brokerage houses. The place that serves as a background could be anything else as well, depending upon the theme. I have, for instance, imagined an avatar of a soldier on a battle field of the medieval times, somewhere in Europe. The background is the terrain of a battle field. Weapons or accessories that the avatar can choose to possess, in this case, could relate to different investment vehicles like stocks, options, bonds and mutual funds, and each weapon or accessory, during the selection phase, would 'educate' the player about its features in a concise blurb. Like the investor's avatar himself, these instruments also can have their own variant-avatars to match with the overall theme of the game. Although the game itself may provide basic information, there could be buttons or links that would lead the prospective investor to a site where you could provide extensive information for investor-education. You may choose to charge a little sum for registering and getting trained, or leave it to be a free and non-commercial offering. The more important solution to look at promoting would be investment advisory services.

- Each weapon or accessory could be 'augmented' based on the points the investor's avatar would win, forging through the game. For example, a stock accessory can be upgraded to a preferred stock accessory based on the number of points won or you may simply choose to reward the player with more lots of shares. So, on the game I have envisioned, a knife (common stock) may take the form of a gilded sword (preferred stock), for example, when it transitions through the augmentation.

- Each strategy in investment could become a different level of the game, starting with simpler strategies and advancing into more complex ones. Again, this would depend upon the theme. Titbits of information could appear on the screen anytime a new term is introduced, and a little puzzle or quiz could be part of the game to enable the player to choose the correct answer and win bonus points. In the case I have picked, my soldier can advance into tougher terrains, from the plains to mountainous regions as the game advances.

- An advanced level of the game could involve letting the user build his own portfolio with his chosen asset classes, and you may let him invest into a mimicked market sphere, so that he can taste virtual returns. Also, there are several investors who do not get a clear picture of their year-end statements. They would be benefited by learning all that goes in, and what fees is applicable and why. This will reinforce the trust factor into your product offerings in a subtle manner. My soldier might possibly build his portfolio so he becomes the army's commander at the end. He would be intimated of all that applies to him as part of his rewards in the elevated position, mimicking the style of a portfolio-return statement.

- Once an investor is familiar with the terms of the business and understands how market conditions have a bearing on his corpus better, it could work as a strong impetus for him to invest. Familiarity, in this case, breeds business!

- People who may be saving money, but are risk averse with investments would certainly become more inclined to make their money get them returns. Therefore, you could offer, through the application, the choice of investing or taking investment advice from your firm. A simple button like "Invest now for real!" or "Talk to your personal advisor now!" would help provide the business interface needed.

The professionals from your company could further help the investor with selecting the right portfolio mix. This would be an interesting way to build a trusting relationship because the usual barriers of sales can be easily broken by interfacing with the customer through the means of a beautiful and innovative app.

2) Personality to portfolios through VR:

It is a common fact that people's personalities impact the products they buy and the choices they make with regards to everything, not just investing. If there is a product that could accurately map personalities to the portfolio mix, it should make a lot of business sense. If Google can give personalized search results, and we can get personalized news feeds based on our interests, then why not a portfolio strategy to suit our personality?

Risk appetite is one of the major factors to consider in building an investment portfolio, and this is directly a part of one's personality. The traits reflect in the actions, and that could be trailed off the smart systems to make intelligent business solutions. It may mean that we would have to invest in big data, AI and data analytics and IoT, but it would be well worth the effort considering the long range of the product's road-map.

An efficient portfolio-building virtual reality application could enable investors to choose from some of these–hobbies, fields of interest, markets of interest, whether they would like blue chip or penny stocks, whether their ideal character of interest would be a cowboy, a technical geek, a philanthropist, etc. to do behavioral and personality segmentation accurately. This would be a far better option than answering routine and perhaps boring questions on the investor's financial goals, income levels, risk appetite, age, etc. These usual traditional questions that are asked in wealth management and portfolio building can still be answered through intelligent systems that can tap into multiple sources without bugging your client with a form to fill in, and select from multiple options on what is applicable to them. This is where we should think of synergizing better with IoT, AI, big data and data analytics solutions. If the investor has already got a trail of previous investments to pick from, it gets easier.

The last step is to make an intelligent mapping between the customer's selections on the virtual reality app, and the information you get out of your analytics–this will become a recipe to build a customized portfolio to suit your client's needs better than by using traditional methods of wealth management and portfolio building.

3) An immersive VR treasure hunt game for wealth management, mapped with real ROI yields:

People love the concept of hunting for treasure, the world over. Treasure though, may be of any form. It could be precious metals or gem stones, hidden information from the past, like that on the dead sea scrolls, antiques including paintings, and so on.

This fascination and 'need' that humanity has for treasure hunting may be understood by looking at the frenzy of the gold rush, the expeditions undertaken to find the Ark of the Covenant that is believed to contain the Ten Commandments that were revealed to Moses, the several trials to recover treasure from shipwrecks like the 'Nuestra Señora de Atocha' etc. (The 'Nuestra Señora de Atocha' was a Spanish ship which was part of a fleet and contained precious metals such as gold and silver, jewels and gems, amongst other cargo when it sank into the deep seas during a bad hurricane in the year 1622. The Spaniards had to borrow heavily to fund the Thirty Years' war they were waging at that time because of the immense loss caused by the ship's sinking. It took until 1985 to finally recover much of the long-lost treasure.) Building an immersive virtual reality underground maze, or a deep-sea diving treasure hunt game to unlock hidden treasure would be an interesting way to pitch sales to the Gen Y and Gen Z people. While the player finds the leads to unearth the treasure, the game could display the actual returns on investments that have been previously made through the products of your firm. For example, I imagine a case where the player has picked up a wooden chest and a leather bag from a pirate ship on the ocean bed. These could be mapped to two different types of portfolios, and each one, when opened, would display the treasure in the form of returns on investment from that particular type of portfolio. This would be appealing in the sense of aesthetics, as well as, will be backed by real data on the value your customers have derived by banking upon your service offerings.

When your pitch is backed by data and statistics, it would automatically drive further inflows into your assets under management.

4) Using the time machine trick and AR to propel goal-based savings:

Data visualization is a very powerful technique to build your business case. Clubbed with augmented reality, it can give astoundingly good results. Because seeing is believing, people who want to 'look' into the future would love if you could put them on a time machine that would take them into their future. You could use the customer's video image and surroundings to give a realistic view of the future him, and build a real-world-like environment in some point of his future to show him what your investment products could empower him to achieve.

Tricking your investor to look into his future can help him assess his monetary needs for the future better, so that he can plan his needs, goals and investments better. At the end of the day, no matter what anyone says, nobody really does live one day at a time. We all look forward, we all think of the future, and most of us do that for more time than we think of the present.

A proud parent beaming at his child during his college graduation ceremony, a loving couple celebrating their retirement days on a beach holiday, a young couple in their dream home, a travel enthusiast touring some of the best travel destinations on earth, these are powerful simulants to trigger great savings and investments. It would not be just about making business; it would be about adding value to lives.

Intelligent data could help you map and port your investor's avatar onto the right template of vision by augmenting his real world–on a beach, a cozy home, whatever! Knowing your client's interests and passions is the key, and showing him what he would like or need for his future is the door for channelizing investments.

5) Copycat investment solutions though VR:

In several spheres, copying is a cheaper and more efficient option to attain results. Investments are not an exception to this. Copycat investing is replicating the investment formula of seasoned and successful investors like Warren Buffet. Many hedge fund and mutual fund managers do follow this strategy even though it has its own associated risks. You as a business firm, can let your well-informed investors choose whose footprints they would like to follow while formulating their investment strategy. Is it Mr. Warren Buffet, is it Mr. Carl Icahn, or is it the S&P 500 index? Whatever the strategy, it will be very beneficial for investors to see the kind of portfolio mix they can choose, and get an anticipated return in exact figures and charts, and clear-cut ROI reports on a virtual reality app.

When the goal of making money is in itself a great enticer for human enterprise, it should be further more so to see a clear-cut representation of the figures, and how the fund manager at your firm allocates the customers' funds to maximize returns. Besides, the charges applicable for copycat investing would be much lower than building a customized portfolio in the literal sense, and the onus of choosing a model for investing could be tilted towards the investor himself, if he would like it that way. The advisor will then become more of a figure head for any supportive help an investor might need on the way, in case you would not provide the benefits a full service broker does. Offer the flexibility to rebalance the portfolio anytime and with lower to no additional fees and it should be more than perfect for many investors. This will put them on top of their investments as well as reduce overheads in investing.

6) The news channel style of MR app:

We all know that serious investors would never shy away from news. Current affairs in everything from economics to politics have a hard-hitting impact on investment returns. In fact, most investors today show an inclination towards being 'in the loop' with regards to the management of their portfolios in tandem with market trends. Therefore, it will be a good idea for you to post your investors with the news that could impact their portfolio returns, and on MR, help them put two and two together by rebalancing their portfolios for the maximum possible gain and benefits in the scenario or market situation that exists to be handled. This approach would help your investor know that their investments are being managed proactively and in the best possible manner under the given circumstances. A proactive business that concretely displays how it is constantly keeping the client's best interests as priority would be certain to go a long way. A very significant example for what might hit with a massive change on the returns on customer investments would be BREXIT, Britain's exit from the EU (European Union).

Finally, it will be well worth to remember that the best business partnership on the planet is the partnership between a trust-worthy business and its client.

7) The investor's 'own' robo-human advisor in MR:

Some people trust individuals more than they would trust firms even if they may be world-class, top of the range business firms. A lot of people actually turn to their buddies for advice with any matter of importance, rather than seeking a professional's help. This applies to almost anything to do with human enterprise, though it may sound a bit surprising.

If an application with technical and business depth can appear like a simple, humble, human being, we could possibly put in more of our trust in it. If you are an advisory firm and want your individual investors to have a trusting relationship with you, you would have to first build a 'human like', and not a 'business like' relationship with them.

Building a mixed reality app where a human advisor and a bot advisor come together into one avatar and creating the interface needed for your individual customer to talk to, can take your business forward in a much better fashion. The talk need not be purely business. There are so many chat applications in the world that can talk to an individual about any topic under the sun while learning intelligently on the way. Build the kind that can talk to your customer on any financial aspect, and alley their fears or discomfort. Ensure that he would get to invest his money only once he is into the required comfort zone, and it will gradually scale up your business. Although relationships in business may seem to have less to do with the loyalty factor, customers would not let go of a business that would sell a product or service with the trust and comfort factor. This is a unique way to show, tell, and get your bargain.

When individuals are 'human', businesses better be 'human' to see that barriers to do business can get broken easily. The point to remember is that advisors are not salesmen. The advisor's goal should be to help their clients with investing into the right product to meet their financial goals, rather than talk the clients into buying the wrong ones to get their sales bonus and get away with it. The world has seen many instances of malpractices due to this sales bonus component of one's compensation; do not wait to show more. Link benefits and compensation to actual client satisfaction factor, rather than the revenue factor. When your clients are made happy, the money comes in automatically. You retain your old ones and gain new customers. Run after the sales targets and numbers though, and your business would be restricted to being only a short-term winner, and a number crunching machine, at best.

If your firm can really be committed to the above point, and add real value to the client, then you can be sure to gain more through a returns-based advisory fee. This means that you let your customer pay you the fee only if he sees his targets materializing, and perhaps, pay more over the base fee if he is happy. That is, add a 'happiness quotient fee'. You might be surprised that there will be people who would be really happy to part with more money than you would ask them for, if only their experience with you had been good.

8) VR investment solutions for the penny-wise too:

It is the little rain drops that trickle down into the channels, towards the seas that make the waves of the tide.

Small investors can bring in huge business too, and you can make innovative solutions to build a fund that they can all partially own and grow their collective fortunes. A simple virtual app can aid small investors to select particular funds they may like to invest in, from a given range of fund options for as little as $5, and be told as to what they can expect to gain on the horizon, along with the applicable market risk conditions. Money made is money made, irrespective of the actual amount of returns.

You may make this a DIY (Do It Yourself) style of fund selection with expert advice available for a decent little sum. The investors may like to be more in control of exactly what they would like to pick for their investment, with a little professional support. Ensuring that there are minimal to no balance requirements should catapult a growth in investments. Sometimes small is the real big, as a huge small investor segment can give you business returns that the giant investors might generally bring in.

9) Give vodka to the alcohol lover and todka to the chocolate lover–customize, and customize upon the customization:

Clients come in a multitude of variants. The more there are, the more their preferences and needs. Within your array of products, offer a wide variety of options for customers to select from. Customers who sense the freedom of doing their business with you are the ones who will stick, since in the end, all of us are lovers of freedom and independence.

A business that can capture the psychology of a human mind would be on auto-pilot when it comes to garnering accolades from its satisfied customer base. The below ideas can be given the needed shape in any of VR/AR/MR as you may deem suitable.

A few ways to enhance the business scope of a product, would be to offer:

- Specialized market intelligence and impact reports for your UHNV clients at a little additional fee–they would not mind if the predictive analysis is good with the usual conditions applied. This implies that you may offer portfolio management services to automatically happen based on not just general news feeds, but also on market intelligence reports that are accessible to the privileged few. The differentiation between the value-add from portfolio management based on such specialized intelligence as against from the basic portfolio management services can be represented such that it is perceived the moment the investor takes a look at the picture.

- Offer a very wide range of investment options, from the conventional stocks and bonds to the more unconventional hedge funds and real estate investments, along with offering accredited investments that are exclusive for the high net worth clientele and manage them all with the backing of predictive analysis. Treat your client's money like your own, and your management of it will come mostly right. People have lost so much to the 2008's downturn only because of the unprecedented risk the investor's money was put into, and due to a complete disregard for the need of a little caution and discretion. Speaking of striking a balance between the risk element and offering unconventional solutions, there is always a middle ground. As an example, the industry has been seeing a steady and healthy spike in the real estate investments in Canada, going by the reports from the Canadian real estate association. This serves as a good non-conventional, but 'safe' option to put in some funds and sustain a healthy growth. Investors who are risk averse but still want to dip their toes to test the waters may start small to grow big through investment options such as these. Modeling these in architectural representations would be a pretty effective way to make the investment vs. returns come alive for the customer. The customer can himself place the building material to form the architectural structure in a solution where it would represent the various investment options, and the size of the structure being built could convey the potential future returns, for example.

- Provide options to drag and drop or move around the visual elements of a report or dashboard to get customized statements. If you take notice, a lot of people stick to a specific newspaper not because of the news they report, but because they got used to a certain presentation style and thematic representation. Otherwise, most news come from common news agency sources like Reuters, and get reported without too many 'touch-ups'.

- Let clients choose whether they would like an actively managed portfolio or whether they would stick with a passively managed one. The fees may be determined based on the frequency of rebalancing the portfolio as well as the market condition at a given point in time. After all, people would not like to stake a mountain to safeguard a molehill. Here too, displaying the difference between the value that can be derived out of an actively managed portfolio vs. a passively managed one through visual representations would be a much helpful elicitation method.

While the mature clientele may, or may not have too much time or interest for playing games or building structures, they would not want to lose at the game that matters. After all, this should matter more than golf or tennis. I would also like to reiterate that it is more than a compelling reason to build solutions like the ones I have presented for the obvious thing: newer generations of people are addicted to electronic devices and games, most of which are seemingly purposeless; but then, we can bring a purpose to these, is it not?

However, in case you would like your sales talk to be more like a traditional business presentation for your clients, here is a great option, as an example:

With a little time spent on an intelligent augmented reality application, you may demonstrate in your boardroom, right on your white board, how a high net worth investor might get his portfolio built, select the way it is managed and how he may keep track of his returns in a single field of vision, with precise elements (think holograms) that would be summoned by your experts, onto the board; just like a magician weaving magic right before the audience would–by a few sweeping gestures of his 'digitized hands'.

10) Do all products have to be customer centric? How about your business enablers–your employees?

You may have recruited the best of the breed of advisors and fund managers to manage all the money, and be customer centric. However, just as a pilot would go through simulation flights as part of his training, the asset managers can go through a lot of investment and market scenario modeling and stress testing through immersive VR, instead of by traditional modes of training.

Real as well as imaginary scenarios can become case studies to ensure better performance levels. Even the best universities cannot teach everything; a lot of learning happens on the job and on the preparedness of the person himself to tackle the tough situations and dynamics of the market as they keep taking shape. You would need to keep training your employees to ensure that they are equipped to build you solutions to stay ahead in the market. Testing of portfolios against events or occurrences impacting the market is done anyway, but it can be done in an interesting and innovative approach using the fourth wave of technologies. As an example, in immersive VR, a stock market crash could come alive in an avalanche over a snow-capped mountain or through a building that comes crumbling down to dust when your fund manager makes all the wrong decisions that would lead the funds under his management to go down in value. This would just reinforce the importance of making the right decisions about the assets under your management because the impact would stay lucidly visible in the mind's eye when people experience immersive VR, rather than when they just go over figures and graphs.

For all the skeptical or conventional thinking people who would say that investing is serious business, one thing to remember is that business and fun do go together; it is just a matter of getting used to that kind of model in doing business. In the financial industry, we seldom experiment, and stick to rules and regulations; this is why the domain lags behind in innovation. Although it may be imperative to stick to the rule book, we must remember that blockchain technology could not have happened unless somebody changed the rules of the game and brought in Bitcoin. While we spend millions in developing products that are not so very unique, and that a lot of competing firms are offering as well, why not go a little out of the way to get some products to be presented in a unique fashion? It just might show a great spurt on the credit side of the balance sheet.

Where the implementation of an idea involves having the necessary gizmos to aid user experience, the potential drawback, as on date, could be the cost factor. The costs for the hardware as also those for building the applications could be high today, with these technologies still in the nascent stages. But this will change sometime soon. Meanwhile there are alternatives.

Your company's sales representatives or client interfacing officers may take some gizmos around for demonstrations, you may set-up some kiosks at your brick and mortar offices for clients to 'experience' your solutions, or you may ship some gadgets to your HNV or UHNV clients to check out in their comfort zone. The play-stores as well, could offer your innovative applications to interested users for download and use on multiple platforms. Impediments anywhere, in many cases, are more due to being risk-averse and exercising the devil's advocate before investing into business.

To grasp, is this–ideas may conform with logic, or may defy it altogether. Ideas may conform with what statistics and intelligent data reveal, or may create them all anew. Ideas may be appealing or they may not even register on one's mind as anything noteworthy. Ideas may sound great, or they may seem pathetic for business-scope. But what is important, is that ideas are our turning points in life.
They are those that set a lot of things in motion, and change the direction of the way the world works.

There is no good or bad idea. It is only good or bad reception of the idea that makes the difference, and the world is not exactly predictable with the fast-changing dynamics of today. Behind every successful human enterprise, is an idea…the idea, is not to kill the idea!

Watching the 'Wave'

The following will serve as a briefing for anyone querying, "Hey, What's up?". Even though estimates and figures may vary, this will leave nobody in doubt that the time is very ripe for the next big thing in technology, the fourth wave.

1) Fidelity Investments has taken the leap into the virtual reality world by developing an application to help drill down into its employees' asset allocations and savings in their retirement accounts, and also check their age, and the number of years they have been serving. The idea is to help their employees improve their savings for retirement, and make better decisions on their investments. They have used StockCity virtual reality application, where the representation of portfolio allocations is done through virtual, 3D buildings that can be experienced through Oculus Rift.

   Is it not that when people see better, they can manage better?

2) Bank of America's online discount unit, Merrill Edge, has brought forth an application called 'Face Retirement'. This one scans your face through the site's camera functionality, and reveals how you would look like in the future. This would help you see two things: the fact that you will be growing old, and more importantly,

that you would need to save for your retired, old self. This is a very powerful cue to invest for the future. Obviously, Merrill Edge stands to gain an edge in the business.

3) MetLife, a pioneer in providing insurance, has launched a 3D simulated virtual reality environment that lets customers interact with their insurance advisor in the form of an avatar. The platform is aimed towards making the customers being able to solve servicing issues, and be the protagonists in their own financial journey.

4) Goldman Sachs identifies virtual and augmented reality as having the potential to become the next big computing platform. As per a research statement from the firm, these technologies are headed towards becoming an $80 billion market by 2025. That means the wheels are turning.

5) Back in 2012, through a Kickstarter campaign, the Oculus Rift headset made $2.5 million for development. Since then, every other player has got onto the bandwagon, including big players like Sony, Google, HTC and Samsung, to develop devices that would put consumers in wonderland.

Today, some of the best VR headsets available are Facebook's Oculus Rift, HTC Vive, Sony PlayStation VR, Samsung Gear VR and Google

Daydream View. Although the technical specifications vary, they have been changing the way the world is being viewed.

Microsoft, meanwhile, is partnering with companies like Lenovo, Acer, Dell and HP and getting less expensive, but very immersive VR headsets on the market, and the more expensive but very enticing mixed reality experience through the HoloLens that blends the users' experience with virtual holograms.

6) The sales of VR headsets are soaring, even though the applications are restricted to gaming for the most part. However, two pointers here are that the millennial and centennial generations are addicted to gaming, and business can be well connected to gaming, as we have seen in the previous chapter.

7) Going by a study made by ESA (the Entertainment Software Association), of the more than 320 million people in the United States, 155 million play video games. About 56% of this population is under the age of 35 years, and 30% is in the age group of 18 to 35 years.

As per a poll by Zogby, 57% of people in the age group of 18 to 34 years play video games at least thrice a week, and a whooping 66% felt that they get to gain an understanding of new technologies useful for life by gaming. Further, more than 28% people of this age cohort disclosed that they have an avatar online, an alternate, virtual them.

As per Newzoo's latest global games market report, the estimated returns from gaming in 2017 is around $108.9 billion.

8) With several companies being secretive about their sales figures, arriving at the actual sales figures of VR/AR/MR devices may be difficult. Also, the potential for future sales growth may have been grossly underestimated, considering that the hardware would keep getting cheaper than it is today, and the numbers would keep adding up from economically and technically developing parts of the world in an exponential fashion. Pundits of the industry are hugely backing the tremendous growth potential of this technology wave.

As of March, 2017, statistics have come up with the following data on the sales and usage of some popular gadgets:

- PlayStation VRs sold over a whooping 915,000 units

- Samsung's Gear VRs users spent over 10 million hours in viewing 360 degree videos
- HTC has sold over 140,000 units of Vive

9) As far as the Oculus Rift is concerned, a lot seems to be happening behind the screen, or is being kept under wraps. However, the scores of thousands of developers registered on the platform, including experts being roped in from reputed animation companies like Pixar, tells much of the story.

HTC is working with entertainment companies like HBO, Lifeliqe and Lionsgate for developing VR content, and Samsung is partnering with the likes of 20th Century Fox.

10) Business Insider anticipates that shipments of VR headsets would see a compounded annual growth of 99% between 2015 and 2020.

11) As per reports from Bloomberg, China's VR market alone is going to get to the tune of $8.5 billion in revenues, and there are a couple of hundred startups in the country going to serve the VR market segment. Many of these companies will produce far cheaper VR gadgets than we have on the market today.

12) International Data Corporation (IDC) predicts that the revenues expected to be generated from the virtual and augmented reality market will cross $162 billions by the year 2020.

13) The Sci-fi game Raw Data, has reported $1 million sales in a span of just one month. Pokémon GO did way better with a revenue of $600 million in just three months, which is more than the entire market for VR games in the year 2016.

14) Statistics show that two out of three people would be interested to see the virtual tour of a business. This makes it an interesting area to garner business through immersive marketing. The art of immersive story-telling is driving business growth to new heights, such that the number of companies helping clients with immersive virtual reality marketing has been increasing over the years.

15) The job industry's new employer-entrant is VR companies who are vying with each other to recruit the best of the breed of 3D designers, UI designers, Unity developers, and people whose creative quotient is huge. There is a spike in the jobs posted on online portals for recruiting talent into the VR technology development arena.

All of the above point to one direction–the way forward is using virtual, augmented and mixed reality to the best possible advantage.

Conclusion

A powerful business story of today should combine all kinesthetic aspects to be appealing to buyers, and this applies to most of the products we consume.

Big data, small data or thick data cannot work unless we have a powerful way to visualize the same. It makes perfect business sense to invest into virtual, augmented and mixed reality for what the future holds, as also for the fact that several companies have already begun to hit at the gold-pot.

While the cost of developing innovative solutions using hi-tech means may be high, and the timeline for a break-even point may not seem to appear at even the horizon today, with the economics of increasing affordability, emerging markets, and decreasing costs of hardware, things would change exponentially and dramatically after a little time elapses.

Imagine the scenario where you can let go of your smart phones for a multi-functional pair of eyewear in the near or far future–smart glasses that will enable you to not just view reality differently using virtual, augmented or mixed reality, but also with the functionality embedded to become your digital assistant, to do anything at your behest with the aid of the simple kinesthetic cues that you provide. A simple tap on the frame, where a tiny processor chip and a microphone will have been embedded, can make them act as your mobile computer. Your tap can trigger the display of 'n' number of choices or menu options that you would see, and be able to easily choose from, through the visual cues you provide by your retinal movements. This way, you will be enabled to make a VOIP (Voice Over Internet Protocol) call or access your email on the cloud. You may also, possibly be able to open satellite maps, check real-time traffic, and get yourself admitted to your choice of a hospital in case an emergency strikes by calling for the hospital's ambulance service to pick you up from your current location–all with just one pair of smart glasses. While the doctor is performing a surgery upon you, you may choose not to be under anesthesia, but still dodge your pain by drowning in an immersive virtual experience and then pay the bill by selling some of your asset holdings online in case you are a little short on cash and credit. For security, you may 'lock' the glasses with your retinal map, and leave them be without worrying about who may access your precious data.

If the above is not dismissed as pure fantasy, it is plain that we are very likely, heading towards this kind of a future and we don't want to make solutions today that don't conform with the 'reality' of tomorrow.

Further, I want to state here that I don't believe in science fiction, because science is not fiction and fiction can become science. Anything that we term as science fiction is just the reality of a future date. Great writers and movie-script writers who have envisioned a fantastic sci-fi scenario have only just prophesied the future; how far into the future, is the only thing that remains to be seen. So for some of the nay-sayers who have told me I am fantasizing or just plainly babbling about innovative solutions, I have only one thing to say–"Wait and watch. But then, be prepared to be left out of the action."

Here are just a few examples of 'prophesies' or science straight out of science fiction books or movies:

1) Online shopping, home delivery and payment via cards:

    'Looking Backward' by Edward Bellamy scored it all right, way back in the year 1887–the author has envisioned a physical store where every product is on display, and customers have only to select what they want and pay by a credit card, without having to exchange paper money. The goods get delivered in one day through a set of tubes. Reminds anyone of Amazon?

2) Smart watches:

   Dick Tracy's '2-Way Wrist Radio' features in comic strips of the year 1946. These talk about a watch that is phenomenally close to the functioning of a smart watch, a time piece cum radio phone that can transmit and receive. Sounds like Apple's smart watch?

3) Space travel and the moon landing mission:

   Jules Verne seems to have hit the nail well with his mission to the moon in the book 'From The Earth to the Moon', published in the year 1865. Three men were 'shot' to our satellite on a projectile, and very surprisingly, the calculations that Verne made for the amount of force required to propel the machine have really come close to reality when the actual moon landing happened in the year 1969.

4) Smart eye-wear:

   The movie 'Back to the Future Part II', that came in the year 1989, has talked of smart eye wear that enabled notifications of phone calls–a distant cousin of Google glass, Oculus Rift or HTC Vive?

5) Tablets:

   'A Space Odyssey', 2001's most talked about sci-fi movie, seems to have 'unveiled' the iPad in the past. The movie has filmed devices that look strikingly similar to today's tablets, whereas back then, we were still holding our bulky cell phones.

6) Driver-less automobiles:

   1990's 'Total Recall' has first envisioned cars that were not driven by humans, but robotic drivers. Today, it is close to the reality of Google's driver-less cars.

Could there be any more doubts that the so-called science fiction is mere fiction? Sometimes, fiction that appears to be fun is pretty serious business. Happenings akin to what was presented in 'Journey to the Center of the Earth' by Jules Verne and 'The Time Machine' by H. G. Wells may be just around the corner. I am certainly keeping a watch. (And so should you, if you are a real enthusiast at heart).

I would like to repeat this for the sake of emphasis: for all the skeptical or conventional thinking people who would say that investing is serious business, one thing to remember is that business and fun do go together; it is just a matter of getting used to that kind of model in doing business.

In the financial industry, we seldom experiment, and stick to rules and regulations; this is why the domain lags behind in innovation. Although it may be imperative to stick to the rule book, we must remember that blockchain technology could not have happened unless somebody changed the rules of the game and brought in Bitcoin. While we spend millions in developing products that are not so very unique, and that a lot of competing firms are offering as well, why not go a little out of the way to get some products to be presented in a unique fashion? It just might show a great spurt on the credit side of the balance sheet.

With all the data points that we have today, it just seems that virtual, augmented and mixed reality are exactly the need of the hour. These technologies will give the much-needed shape, size, color, and investment appeal to our products.